SpacePort
AUSTRALIA

www.spaceportaustralia.com.au

Dr Gabrielle M Caswell
BSc (NTU), BSc (EDU) MB;BS (UQ), BA (UNE)
Dip. Pract. Dermatology (Cardiff University, Wales UK)
Master of Medicine (Primary Care Skin Cancer Medicine) (UQ)
Master of Arts (Trinity Saint David College) (Wales, UK)
Master of History (University of New England, UNE AUST)
FRACGP, FACAM, FACRRM

Publisher
My Space 1-10 ISBN 978-1-7635512-3-7

www.spaceportaustralia.com.au

PO BOX 705
Moree NSW 2400
Australia

My Space ABC

Early Reader 3-8 years

This book belongs to:

....

Dr Gabrielle M Caswell

Aa
is for Astronaut

Astronauts train for a long time to go into space, they are lucky to be able to see Earth from space.

Bb
is for Bolide

Bolide are meteors that explode when they reach the Earth's outer atmosphere and burn very brightly.

A is the Alien's favorite letter

Cc
is for Ceres

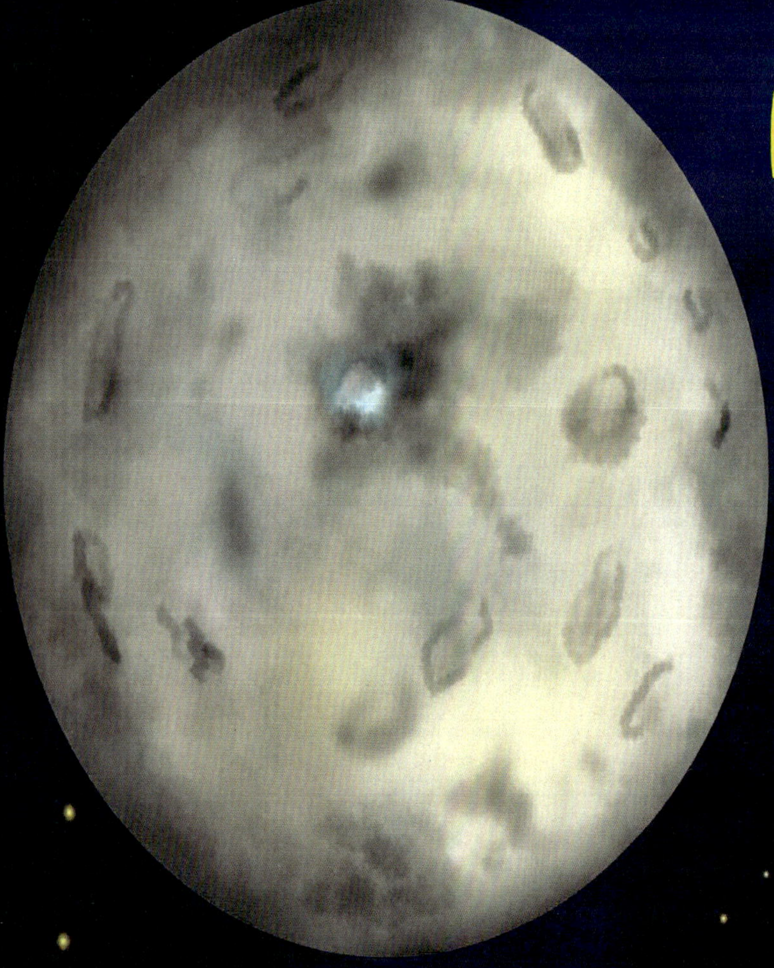

Ceres was discovered on 1 January 1801 by Giuseppe Piazzi at Palermo Astronomical Observatory in Sicily. It is the largest and only dwarf planet in the inner solar system, it can be found in the middle of the asteroid belt, sitting between Mars and Jupiter.

Dd
is for Draco

Draco is a constellation in the far northern sky. Its name is Latin for dragon. It was one of the 48 constellations listed by the 2nd century Greek astronomer Ptolemy, and remains one of the 88 modern constellations.

Ee

is for Earth

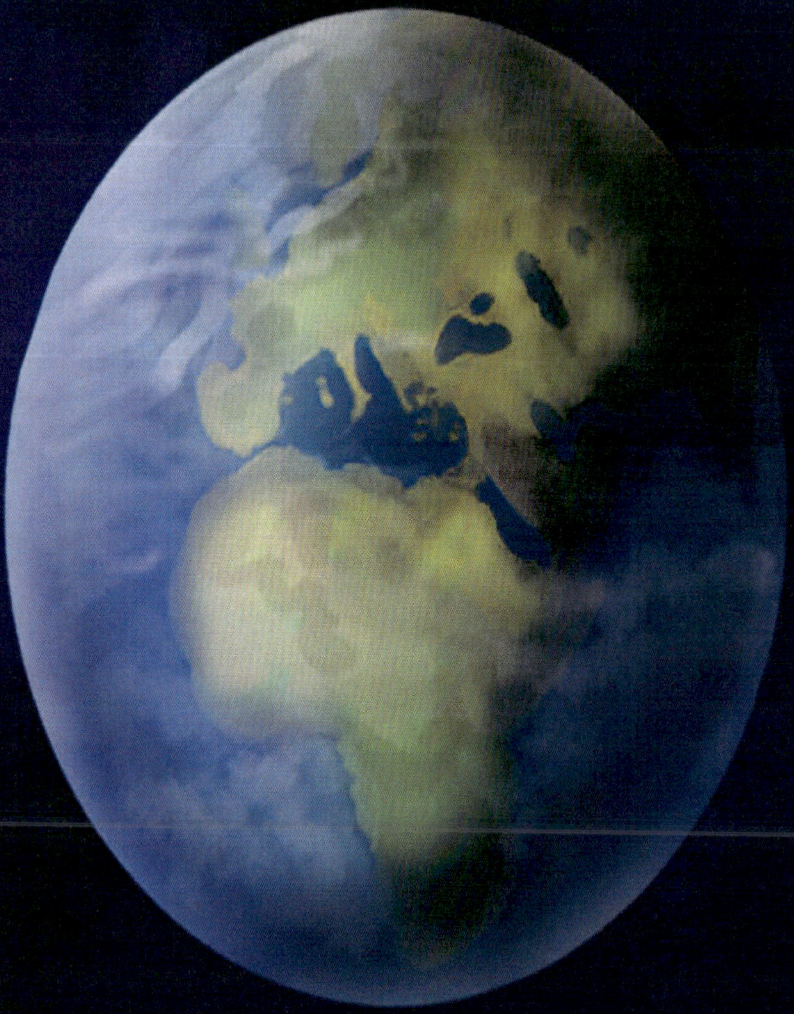

Earth is human's home planet, it is third from the sun, and it has oceans that cover 70% of its surface and support sea life. The Earth has one Moon and is the only planet in our solar system that human's know has life.

Ff
is for
Falling Star

A falling star is a meteor which can be seen when it moves through Earth's atmosphere. Although a meteor may seem to be a few thousand metres from the Earth, when they enter the Earth's mesosphere, at around from 76 to 100 km from the Earth' surface, they usually begin to burn brightly.

Gg
is for Galaxy

A galaxy is a large system of stars, interstellar gas, dust, and dark matter, bound together by gravity. The word galaxy comes from the Greek language: galaxias, which means 'milky', our galaxy, with contains our solar system is called the Milky Way.

All the Letters

Hh

is for Helium

2

He

Helium

Helium is a element that has the atomic number 2, meaning it has two protons in its nucleus. The original meaning of the word helium is from the Greek language and means Helios, the 'sun'.
The chemical symbol for Helium is He and its atomic number 2.

Ii

is for Io

Io is the innermost and second-smallest of the four moons of the planet Jupiter. Io is slightly larger than Earth's moon, and belongs to a group of four moons, called the Galilean moons. Io is the fourth-largest moon in our Solar System.

Jj

is for Jupiter

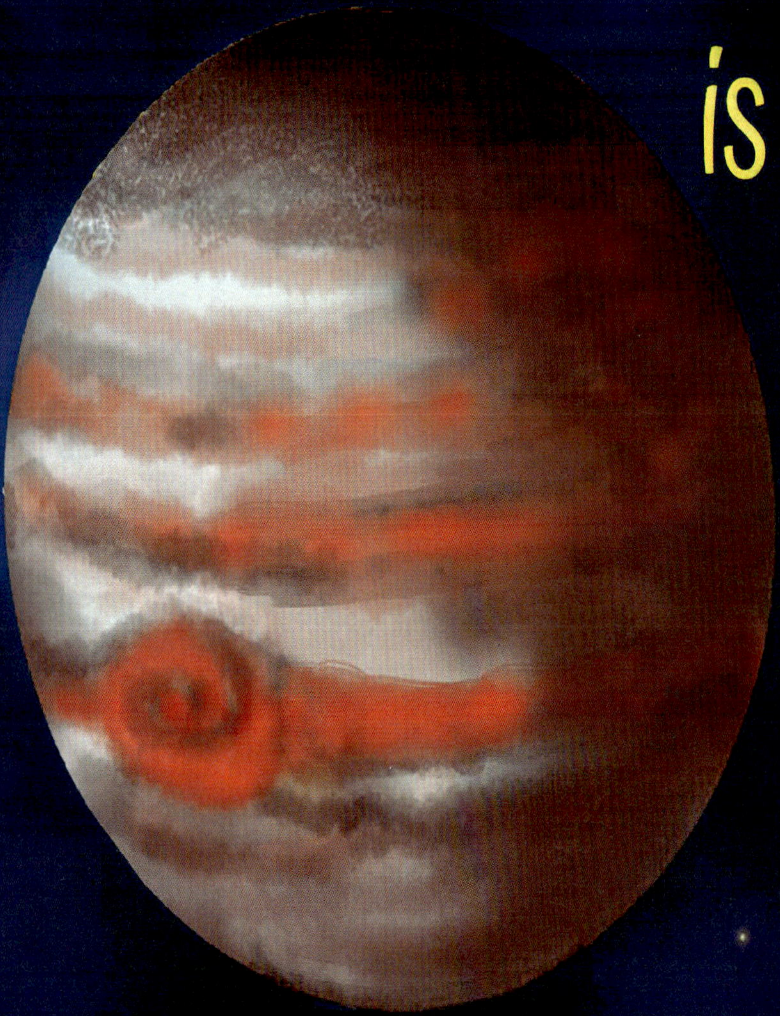

Jupiter is the 5th celestial object from the sun; it is 11 times bigger than Earth. Its atmosphere continually creates the largest storms. Jupiter has 67 moons and is known for its giant storm that looks like a red spot. The red spot is bigger than the Earth.

Kk
is for Kuiper Belt

The Kuiper belt is named after the Dutch astronomer Gerard Kuiper, who deduced the existence of the belt in 1951. The Kuiper belt looks like a large disc surrounding the outer Solar System, reaching out from the orbit of Neptune, and made up of many objects.

Ll
is for Lunar

Lunar means relating to Earth's moon. Our moon is our closest celestial body and one that will play a very important part in space travel. They will establish the Lunar Gateway there, as a stopover when humans travel to Mars.

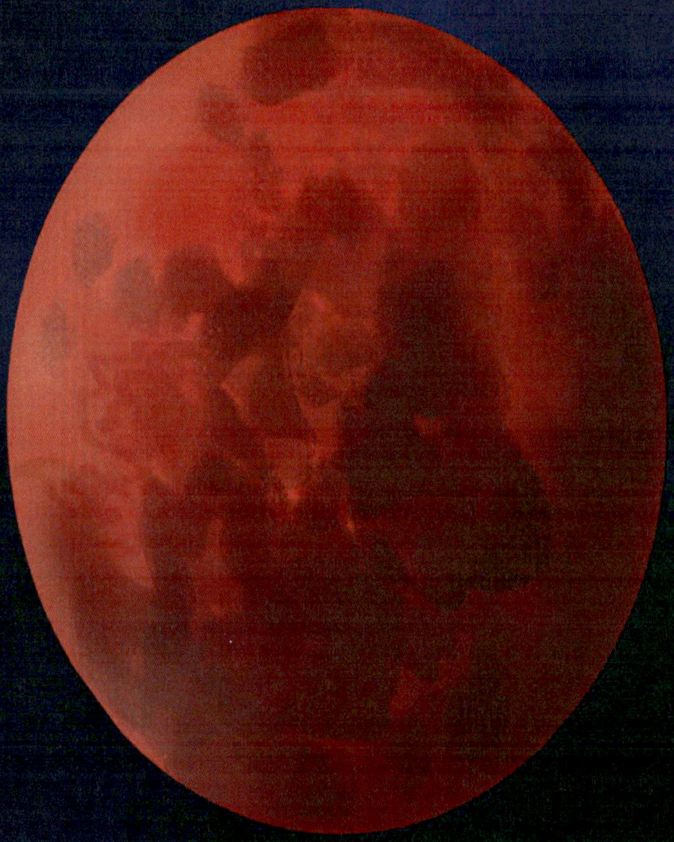

Mm
is for Mars

Mars is Earth's nearest planet neighbour, and it is called the Red Planet because it looks red coloured through a telescope. Just like Earth, it has seasons, this is due to a similar tilt when it spins around on its axis and then moves around the sun. And like Earth, Mars has volcanoes, ice capes at the polar regions, and deserts. The temperature is very cold as it is further from the sun than Earth is. Mars has two small moons, they are called Phobos and Deimos. They are named from Greek mythology, Phobos and Deimos' father was Ares, who was also known as Mars.

Aa
Astronaut
Bb
Bolide
Cc
Ceres
Dd
Draco
Ee
Earth
Ff
Falling Star
Gg
Galaxy
2
He
Helium
Hh
Helium
Ii
Io
Jj
Jupiter
Kk
Kuiper Belt
Ll
Lunar
Mm
Mars

Nn
Neptune
Oo
Orbit
Pp
Pluto
Qq
Qasar
Rr
Rocket
Ss
Saturn
Tt
Telescope
Uu
Uranus
Vv
Venus
Ww
Wormhole
Xx
Xray
Yy
Yildun
Taurus
Aries
Pisces
Aquarius
Gemini
Capricornus
Cancer
Leo
Virgo
Libra
Scorpius
Ophiuchus
Sagittarius
Zz
Zodiac

Nn

is for Neptune

Neptune is the planet in our solar system that is farthest from the Sun. It is classified as a gas giant, and is very cold. Neptune has 13 confirmed moons and six rings. You need a telescope to see Neptune and its rings.

Oo

is for Orbit

An orbit is the curved rounded pathway, that something like a planet would travel around a star, our sun is a star and the planet Earth travels around it. The study of orbits is is a special science called celestial mechanics.

Pp
is for Pluto

Pluto is a dwarf planet in the Kuiper Belt.
The Kuiper Belt is the name given
to a ring of bodies beyond the orbit of Neptune.

P
is for Planets

Qq is for Quasar

The term quasar, a made up word originating from putting two words together: "quasi-stellar" which means star-like. Quasar's were first identified during the 1950's, they were identified in photographic images at visible wavelengths. They often look like very faint, star-like, twinkling points of light.

Rr
is for Rocket

A rocket is a vehicle that uses jet propulsion to go upwards. A rocket engine produces thrust by chemical or thermal reaction, the exhaust is expelled at high speed out of the bottom of the rocket, this pushes the rocket upwards. Rocket fuel, or propellant, is carried in its tanks. The word rocket was orginally Italian, *rocchetto.*

Ss
is for Saturn

Saturn is the 6th planet from the sun, and it is known for the seven rings that surround it. Saturn, like the planet Jupiter, is also a gas giant and it has 53 moons.

S
is for Stars

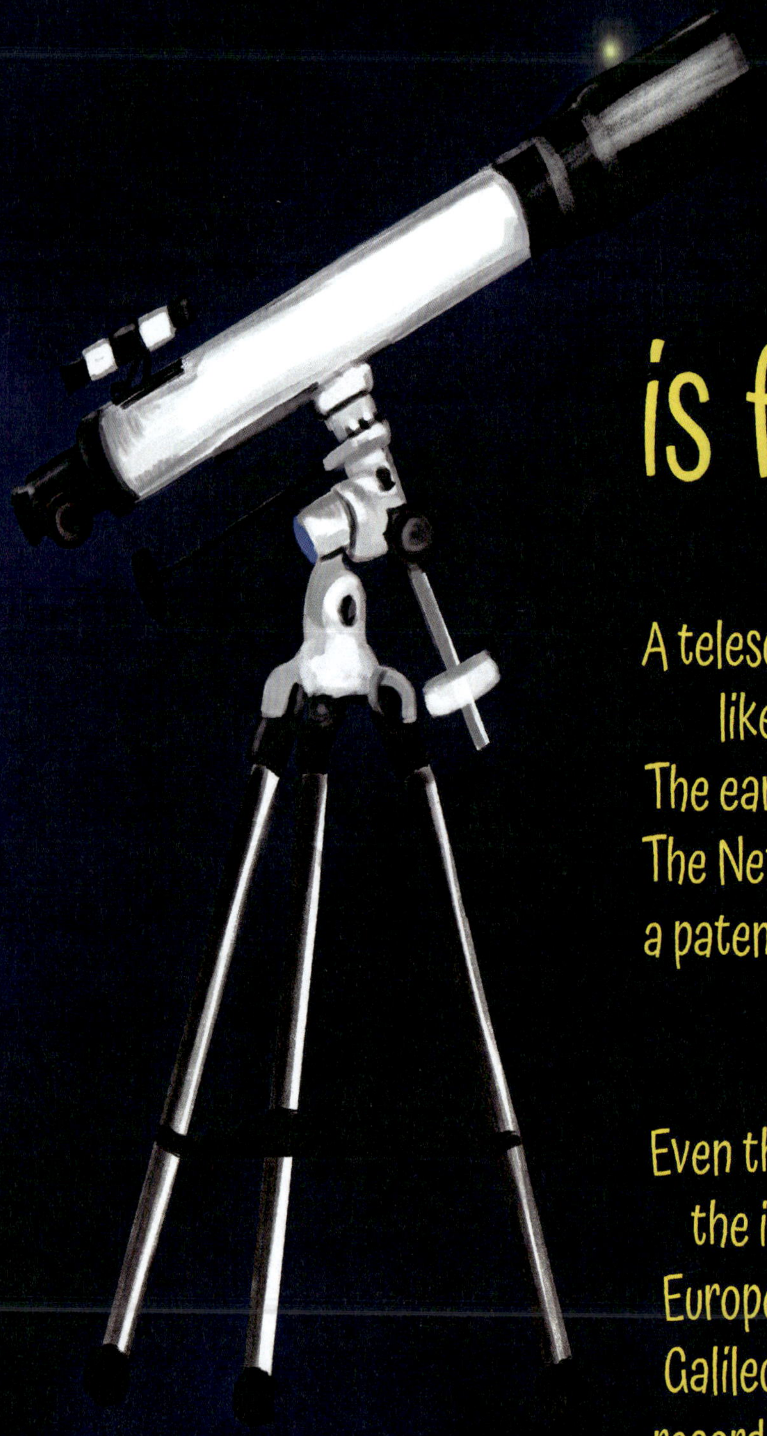

Tt
is for Telescope

A telescope is a device used to observe objects like stars and planets in the night sky. The earliest record of a telescope was in 1608. The Netherlands government have a record of a patent by glasses maker Hans Lipperhey, for a 'refracting telescope'.

Even though the actual inventor is not known, the idea of the telescope spread through Europe, with Galileo hearing about it in 1609. Galileo built his own telescope and used it to record his observations of planets, starts and other celestial objects.

Uu

is for Uranus

The planet Uranus is a gas giant so cold that it is sometimes called an ice giant.
The ice on Uranus is made up from methane, and looks blue.
One year (the time for Uranus to travel around the sun once) is equal to 84 Earth years!

Vv
is for Venus

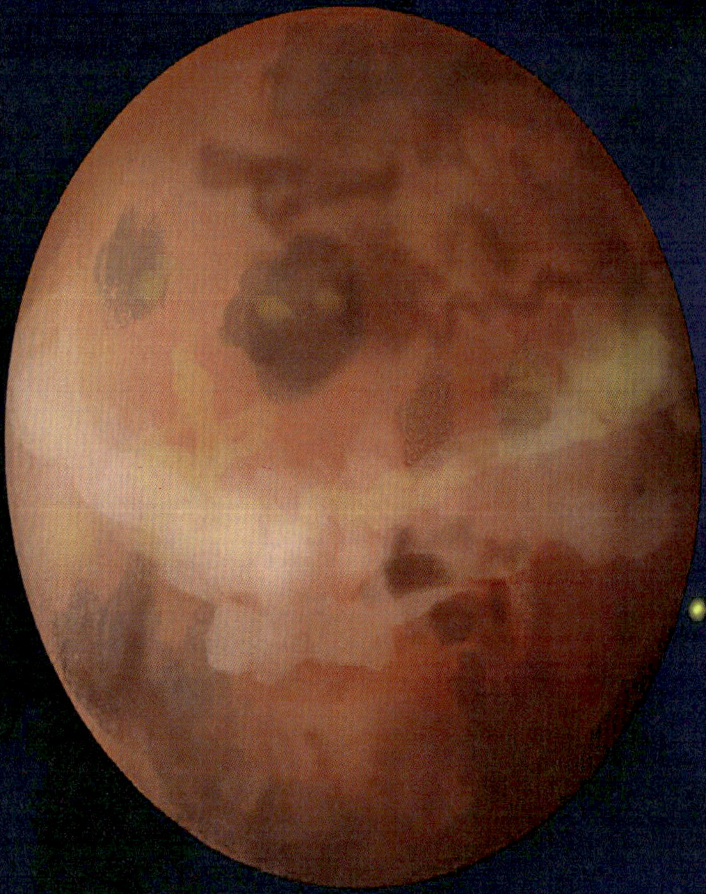

Venus is the second planet closest to Earth's sun in the solar system; it is almost as big as Earth, being around 81.5% Earth's size. It has a thick atmosphere and it spins backwards, the sun on Venus rises in the West and sets in the East. It is the solar system's hottest planet.

Ww

is for Wormhole

A wormhole is a tunnel where the two ends are at separate points in space-time. German mathematician, philosopher and theoretical physicist Hermann Weyl proposed a wormhole hypothesis almost a century ago, in 1928!

Xx

is for Xray

X-Rays were once called X-radiation. When they where discovered by the German scientist Wilhelm Röntgen in 1895, it was not known what type of radiation it was. Science now knows that X-Rays are a high-engery form of electromagnetic radiation. In medicine, X-Rays are used to help doctors diagnose illnesses when people are unwell.

A
B
C
D
E
F
G
H

J
K
L
M
N
O
P
Q

R
S
T
U
V
X
Y
Z

Zz
is for Zodiac

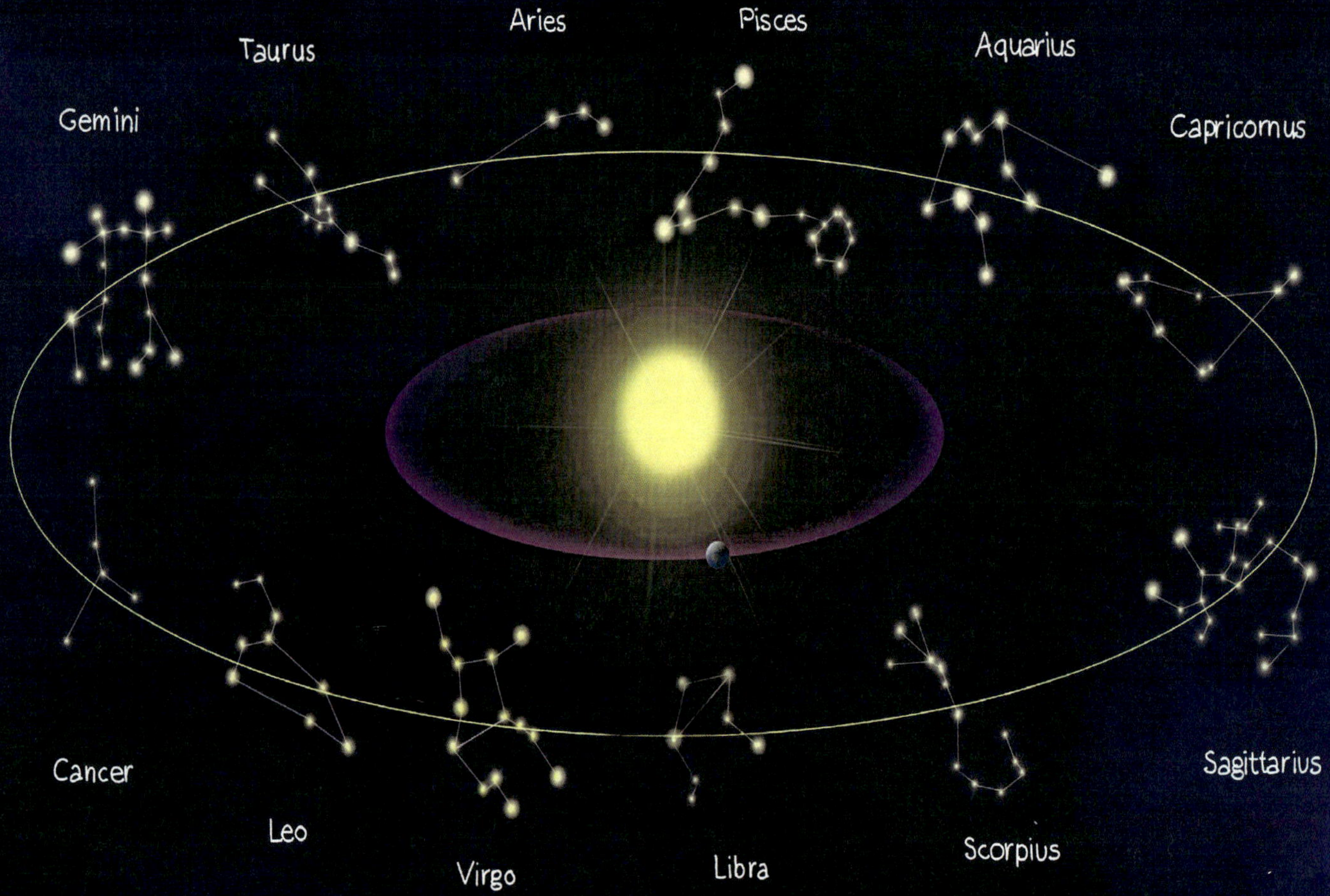

The Zodiac is a collection of 12 constellations that sit around the planet Earth. In ancient times they had been used to identify what part of the solar year the planet is in by observing where the sun was on the ecliptic (an imagined ring around the Earth) and which one of the 12 constelaltions was nearby. The 12 constellations became associated with a month of the year, enabling the Zodiac to act as a yearly clock.

The word Zodiac comes from the Greek word for 'little animals'.

Yy
is for Yildun

Yildun is a star in the constellation of Ursa Minor.
It has another name, Delta Ursae Minoris, and in Turkish this star is call Yildiz.

SpacePort
AUSTRALIA

www.spaceportaustralia.com.au